Is Our Universe Expanding to Form Something?: The Polymerizing Universe Hypothesis

By Malika Ammam, PhD

Copyright© 2018 Malika Ammam. All rights reserved.

Summary

Our current telescopes can perceive the Universe with only an oval shape. What if there is more? This short note proposes that the Universe is polymerizing to form something at the end of the process. It is written in a simplified language that can be understood by nonscientists and the general public.

Have you ever wonder what is going on in our immense Universe? What if we zoom out into a full picture, does the Universe has a shape or is just oval as perceived by telescopes? What if we dig deeper into the seen and unseen regions of the Universe? Well, this could be beyond the means of our current technologies. Even with extremely magnified telescopes, one has to be located outside the Universe to have a correct view.

The Universe is all that surrounds us, from the smallest subatomic particle to the greatest celestial body ever discovered in the cosmos. It consists of billions of galaxies, forming collections of millions of stars with all the space, galactic dust, and black holes. Our galaxy, known as Milky Way, brings together all visible stars at night plus the unseen ones[1]. Besides stars, galaxies involve black holes located in regions of space with high gravity where no matter or energy could escape.

Each galaxy comprises hundreds of billions of stars, forming the luminous bodies responsible for light spots in the sky when viewed by naked eyes or telescopes. Stars are held within galaxies by gravitational forces just like galaxies are held within clusters. The luminosity of stars is generated by nuclear-chemical reactions stemming from radioactive gaseous constituents, such as hydrogen and helium. The sun is the closest star to our planet and without it there will be no daylight and therefore no liquid water and no life. The sun is a giant ball of superhot gas with a diameter around 1.4 million kilometers, which is 109 folds the Earth's width[2]. The sun has a mass of 2 thousand-billion-billion-billion kilograms, which is 330 000 folds that of Earth's mass, meaning that approximately 1.3 million Earths would fit inside the Sun.

The Sun is the center of our solar system orbited by eight major planets[3]. The closest planets, Mercury, Venus, Earth and Mars, are relatively modest in size and rocky in content. The region beyond Mars, known as the asteroid belt, is inhabited by millions of rocky objects, which are leftovers after formation of the planets around 4.5 billion years ago. On the distant side of the asteroid belt exist four gas planets, Jupiter, Saturn, Uranus and Neptune, mostly made of hydrogen and helium gases. The sizes of these planets are much considerable than Earth's dimension but very light for their sizes. Until recently, the most remote known planet is Pluto, an icy rocky celestial body discovered in the 1930th[4]. These planets rotate around the sun in distinct trajectories taking short or long time periods. For instance, Earth takes roughly 365 days to move around the sun while mercury spends less than 100 days and Pluto requires about 250 years to complete the cycle.

The rotation of each planet around the sun is controlled by gravitational attractive forces[5]. The Kepler's laws point that the more the celestial body is massive, the higher it attracts smaller objects towards it. This justifies why every object on Earth's surface is drawn to it because of its immense body compared to the weight of any object on it. Similarly, the sun is a gigantic body which brings planets to revolve around it. Other stars have planets as well rotating around them either in our Milky Way or other galaxies.

More compelling, over 80% of all stars consist of multiple stars systems, accommodating two or more stars[6]. Common multiple star systems involve two stars, known as binary star systems. However, higher multiplicities also exist ranging from arrangements of three to more than seven stars. These multiple star systems are kept closely by gravitational attractive forces to establish either stable distributions with no interactive trajectories or unstable combinations with chaotic behavior due to the strong interactions. The development of these systems is not yet fully understood but some propose that they arise from segregation of collapsing cloud of gas into two or more smaller clouds, later creating multiple star systems.

So, how did all of this begin? Of all proposed cosmological models, the Big Bang is the most accepted theory describing the powerful but remarkable event that started it all. Before the big bang, all space existed as a tiny spot less than a subatomic particle. An explosion then occurred to progressively give birth to celestial objects[7]. During the inflationary epoch, a dramatic accelerating expansion of the very early Universe from the center and outward took place since the Big Bang.

The concept of birth of the Universe and cosmic expansion originated from the observations made by Sir Georges Lemaître in 1927. This information allowed tracking back time of the Universe to one single moment in the past[8]. In 1964, the cosmic microwave background radiation, depicting the big afterglow formed after the birth of the Universe, was discovered, pushing deeper the ideas of the Big Bang[9]. The early occurred explosions, known as supernovas, that formed our Universe and its accelerating expansion, have been associated with the invisible dark energy, an unknown force involved in everything inside the Universe that fed its accelerating expansion.

So what could have been the status of the Universe in its very early life? Well, a high density and temperature state beyond imagination. The early Universe formed 13.8 billion years ago, starting by the Planck epoch era where it could be downsized to tiny packets of primal particles with remarkably high energies[10,11,12]. At this stage, the Universe was smaller than the size of a single particle in an atom at temperatures reaching trillions and trillions of folds than that of molten metal. One important event that occurred during this era is the rapid cosmic inflation induced by strong and electroweak interactions existing in the very early Universe, leading to its rapid expansion from nearly the size of the smaller subatomic particle to that of grapefruit. The tiny ripples formed in the Universe at this stage are believed to produce large-scale structures formed much later.

Just after the exponential inflation, the Universe slowed down to the normal Hubble Rate, around 70 kilometers per second[13]. This early Universe that took about 150 million to develop was loaded with radiation and elementary particles identified in atoms. As the Universe continued to cool down and the temperature dropped below 10 billion Kelvin, new particles were born out of pre-existing ones, including hydrogen and helium. As the

Universe cooled further, negatively charged electrons started to be captured by positively charged ions forming electrically neutral atoms. This process, known as recombination, took place relatively fast. The first fragments of the structure began to assemble, and small clusters of matter increased in size as their gravities attracted other nearby matter[12]. In approximately 400,000 years, the usual forces and elementary particles surfaced but the Universe was still too hot to either produce neutral atoms and other structures or to allow photons to travel fast. Hence, the Universe existed in opaque plasma state. From 380,000 to about 150 million years, the Universe transited to dark ages with no formed large-scale structures. No significant source of light was available at that time, and temperature of the Universe was compatible with that of liquid water (273 - 373 Kelvin).

The dark ages evolved as large-scale structures like stars, planets, and galaxies progressively developed[10]. Compared to Main Sequence stars seen today, stars formed in the early Universe, known as Population III stars, were much larger in size, non-metallic in content, and have extremely short lifetimes. They hence quickly burn their hydrogen fuel and burst as supernovas after only millions of years, spreading heavier elements throughout the Universe. The thin disk of our Milky Way galaxy started to come to shape around 5 billion years ago and the solar system about 9.2 billion years. The earliest traces of life on Earth originated 3.5 billion years ago. This time period is regarded as the most understood but scientific knowledge before that era remains uncertain.

Now, what if we zoom into a much wider picture to look at the smallest of the small? Have you ever questioned what is going on inside every object in the Universe at the infinitesimal level? What would be the tiniest imaginable particle actually look like? The early concept of a possible existing tiny particle was proposed by Leucippus and Democritus[14]. They called it "ATOM" but without advanced technologies, no scientific evidence could confirm this idea. It is until the 19th century that scientists were capable of taking power over these tiny particles to extract energy in the form of nuclear power.

All matter, namely solid, liquid, gas, and plasma is made of atoms either neutral or in an ionized state. These particles have sizes ranging from 62 to 520 picometers (1 picometer is equivalent to dividing 1 meter by 1 trillion)! There is no doubt that you already heard the term "quantum physics", the principles that led to the manufacturing of the atomic bomb that destroyed two large cities in Japan in

1945 and ended the Second World War. That is how majestic the energy that can be pulled out from an atom by disrupting its composition.

So, what makes these infinitesimal particles so astounding? Each atom is composed of a nucleus filled with two even smaller subatomic particles (protons and neutrons), known as nucleons. The nucleus is surrounded by other subatomic particles called electrons, revolving around it just like planets encircling the sun in our solar system. Electrons are negatively charged, protons positively charged and neutrons have no charge, hence neutral. About 99.94% of the atom's mass is found in its nucleus just like the sun compared to its planets[15].

There are two forces at work inside an atom. The first is the electromagnetic force which keeps the negatively charged electrons attracted to the positively charged protons of the nucleus. The second is the nuclear force which is much stronger than the former and responsible for maintaining both protons and neutrons together[16]. The nuclear force causes the protons to repel one another and removal of nucleons from the nucleus will give birth to new elements. Tampering with these forces may generate extraordinary amounts of energy, tackled under nuclear fusion and fission, which can be harvested for either beneficial use like the production of power from nuclear power plants or harmful purposes like atomic weapons.

In the physical world, electrons play key roles in fields of electricity, magnetism, thermal conductivity, and chemistry. They also contribute to interactions in gravity and electromagnetism in the Universe. In 1838, Richard Laming suspected that an indivisible quantity of electric charge may be responsible for the chemical properties of atoms[17]. The term ''ELECTRON'' officially came to light in 1891 by George Johnstone Stoney, and later recognized as a ''PARTICLE'' in 1897 by J.J. Thomson and his group. Inside stars, electrons are called beta particles and are responsible for the nuclear processes and production of radioactive isotopes, known as beta decay. In these reactions, electrons are generated or released to produce power and since they are regarded as particles, they may also be generated during high-energy collisions like cosmic rays penetrating our atmosphere. Electrons with opposite electrical charges are known as positrons, and collisions between electrons and positrons release gamma-ray photons.

So how do these electrons assume all these functions? Well, each electron is surrounded by an electric field, which, in turn, generates a magnetic field comparable to when two magnets face each other. This makes each electron to spin around itself similar to planet Earth around its axis. Electrons are subatomic particles just like neutrons and protons but if their movements are sped up, they tend to radiate, such as when trapped by electromagnetic fields. In other words, electrons carry attributes of both particle and wave. They can be disturbed to speed up or slow down when subjected to external energy. This principle is exploited to harness electrons for good use in electronics, electron microscopes, radiation therapy, lasers, and particle accelerators.

In chemistry, atoms could donate, accept or share electrons with other atoms to form molecules held together by chemical bonds[18]. Molecules may consist of two or more atoms of the same or different chemical elements just like multiple star systems. Molecules consisting of the same elements, such as oxygen (O_2) and hydrogen (H_2), are called homonuclear. By comparison, molecules composed of two or more different elements, such as water (H_2O), glucose sugar ($C_6H_{12}O_6$), and methane (CH_4), are named heteronuclear. These molecules are building blocks of organic and inorganic materials forming our oceans, atmosphere, and Earth's solid components.

The molecules are assembled by two major chemical bond categories: covalent and ionic[19]. Covalent bonds form when atoms share their electrons to produce electron pairs. In this case, each atom provides one electron and both are connected to each other by the pair of electrons. However, each atom maintains its electron pretty much on its side to yield stable bonding that is tough to break, such as in water (H_2O), oxygen (O_2), and hydrogen (H_2). In ionic bonding, however, electrostatic attractions between two atoms of opposite charges take place to create bonds, where one atom pretty much donates its electron and the other accepts it to establish weak bonding. This is what happens in table salt (NaCl), where sodium (Na) with weak electronegativity almost loses it electron and chloride with strong electronegativity mostly gains the electron to form a fragile ionic bonding. Therefore, simple hydration of NaCl in water breaks the ionic bond to form an ionic solution consisting of free ions (Na^+ and Cl^-).

Other types of bonding also exist, such as hydrogen and van der Waals, which keep molecules together to establish larger molecular networks reassembling what look like galaxy clusters[19]. These types of bonds have a significant contribution to the properties of final products. For instance, strong hydrogen bonds are responsible for holding together water molecules in liquid state, which become weak in vapor state and allow more freedom in motion of the water molecules. When it comes to length and angle, molecules are quite fixed by equilibrium geometries, the space as to which they oscillate, vibrate, and rotate along just like stable multiple star systems.

Unlike star clusters which can be looked at by naked eyes, most molecules are extremely small to see but macromolecules could reach exceptionally macroscopic sizes. These include DNA and polymers that may reach from few angstroms to several dozen of angstroms, equivalent to one billionth of a meter. These huge molecules are built by POLYMERIZATION[20]. In this process, monomer molecules like pyrrole exposed to certain types of energy, such as irradiation or electric current, will start polymerizing and expanding in size to create polymeric chains with two or three-dimensional networks. Polymerization may be one principle but the emerging polymers depend on nature of the reacting monomer, inherent steric effects and complexity of the process, which will yield either plastics or biological materials.

Polymers developed naturally through biological processes in living organisms are called biopolymers[21,22]. They could be linked by various types of bonding, including covalent and ionic, as well as electrostatic-like hydrogen bonding which maintains together the three-dimensional structures of larger molecules. One prominent group of biopolymers consists of polynucleotides made of 13 monomers or more, known nucleotides. Examples include RNA and DNA, which form our genetic code. Another group is polypeptides, which are linear chains of amino acids connected by covalent peptide bonds. The third class is polysaccharides, which are long bonded polymers assembled by carbohydrate units. Other categories of biopolymers are rubber, suberin, melanin, and lignin. The most abundant biopolymer on Earth is cellulose that forms the hardest exoskeleton of rigid structures of plants. It is comprised of carbohydrates, and more than 30 percent of all plant matter is made of cellulose. For example, cotton and wood contain 90 and 50 percent cellulose, respectively.

The diversity in structures of the repeated units (or monomers) in biopolymers induces distinct materials, where the biological function is usually determined by the structures and molecular arrangements. For example, amino acids in proteins form polypeptides, which later go through protein folding from primary, secondary, until the tertiary structure. The folding state takes place according to "protein folding", which, in turn, establishes the biological function[23]. These biopolymers are synthesized inside living organisms, originally formed by synthetic chemical processes from basic chemical elements present on Earth or perhaps brought from the outer space as some suggest.

If knowledge referring to the development of biopolymers is still limited, the mechanisms controlling the arrangement and expansion of synthetic polymers are pretty much understood. Polymers are the backbones of plastics, ultimately encountered everywhere, including in kitchen utensil, household objects, work materials, food and so on. The microstructural features and architectural design of synthetic polymers can be complicated than you realize[24]. In addition to simple linear chains, polymers could be grown to form branched polymers from mother chains. This forms star-, comb-, brush-, dendronized-, ladder-, and dendrimers-like polymers. There are even two-dimensional polymers built of repeat units configured in plane-like shape. So, what could be the purpose of acquiring polymers with various patterns and arrangements? Well, they are most convenient when it comes to their physicochemical characteristics like viscosity, solubility in different solvents, transition temperature, and toughness, among others. These features influence the structures and properties of the comprehensive products made of these polymers, such as appearance, color, and durability.

Synthetic polymers may be constructed using single or multiple subunit monomers repeated over and over to generate macromolecules. Various approaches are applied to synthesize stylish polymers. Polypyrrole, for instance, could be assembled from aqueous solutions of pyrrole monomers exposed to energy, such as electric currents[25,26]. Few drops of pyrrole solution are dissolved in a buffer solution to give a solution with identical visual characteristics as those of the buffer solution itself. Upon application of a voltage, polypyrrole will start to grow and both length and mass of the obtained polypyrrole will depend on time of the polymerization and experimental conditions. This approach could be adapted to customize conducting bodies with thin layers of polymers or simply produce polymers in solution for diverse purposes.

The structure and size of the monomers remarkably influence the properties and strength of the emerging polymer. If the configuration of the monomer is altered, the polymer's properties will be changed as well either by improving or degrading. This process could be adapted to form polymers with desirable lengths and weights, and hence better characteristics and properties. Using this concept, polymers with abilities to conduct electricity or emit light (electroluminescence) when exposed to electric currents, could be synthesized, such as some categories of polypyrrole. The degree of conductivity or electroluminescence (emitting light) could be modulated by doping, a process of intentionally introducing impurities into the polymer's structure.

Thanks to science and technology, we all now know that atoms and molecules are building blocks of all matter, which are made of electrons, neutrons, and protons. Atoms and molecules, including polymers, are occupied by 99% void. Equally, the Universe is made of galaxies which contain stars, planets, and many other celestial objects. Most of our Universe is also composed of void, mainly occupied by dark matter and dark energy. So, what if the Universe is polymerizing to shape something at the end of the process? If so, what would be the silhouette of the Universe at end of times?

One way of assembling polypyrrole is by exposing pyrrole monomer dissolved in a transparent buffer solution to an electric current[25,26]. This will initiate the production of pyrrole radicals allowing the polymer to expand. The length of the monomer becomes negligible to that of the polymer as it continues to grow. Likewise, prior the big bang event, the Universe may consist of only dark matter and dark energy homogenously dispersed in what seemed "there was nothing"[27,28]. The polymerization process then commenced with the big bang as the dark matter was subjected to dark energy under proper conditions allowing the polymerization process to begin then expand in size just like a polymer during polymerization. If so, will the Universe reach a definite final shape?

Actually, this will depend on the polymerization degree and the amounts of dark matter and dark energy left to convert into stars, planets, and other celestial objects. On Earth, every polymerization seems to come to an end, where plants, animals and any other lifeforms, reach final growing states before start aging at some point. Indeed, our Universe is over billions of lightyears in distance and yet the energy it contains is still finite. The second law of thermodynamics predicts that the Universe is continuously heading to a state of total entropy, an event associated with disorder and chaos at end of times when the Universe exhausts all its energy as it increasingly expands[29]. This has been speculated since the 19th century as thermal energy is found homogenously dispersed throughout space and no sources are available for its transformation. No conversion of energy means the Universe is heading to death. The formation of black holes, the dark spots displayed in the center of our Milky Way and other galaxies which trap all energy and matter from their surroundings, may be indicative of such end.

If so, our Universe is now in its final countdown of increasing entropy and gravity plays a big role, according to Roger Penrose[30]. It is gravity which collects all matter to produce stars, and it is gravity that causes their collapse to form black holes. All light and energy surrounding black holes become absorbed with maximum feasible entropy. It may be judged that galaxies look flat but black holes apparently have entropies proportional to their surface areas. When these black holes start expanding, objects trapped by gravity are stuck in planes and flat-like structures. It is thus possible that these black holes are a manifestation of our Universe endpoint as entropy continuous to rise in time. Some suggest that once grabbed by the power of a black hole, energy and matter is permanently trapped. The Hawking radiation[31], however, argues that trapped energy may possibly escape because of quantum gravity. In 2014, Stephen Hawking suggested that black holes don't really exist, considering the large event horizon it overcomes. Well, which proposition is correct is still open to debate but entropy still makes a probable end to our Universe.

Well, is our Universe truly going to an end? So far, there is no solid information supporting this suggestion, and entropy still controversial since the time of Ludwig Boltzmann. The heat death hypothesis proposing an increase in entropy of the Universe and utilization of the thermodynamic laws in the Universe are unreliable for apparently large-scale structures[32]. For sure, entropy contributes to the continuous expansion of the Universe since its birth with the big bang event. However, this should cease at a definite point in time, where no further rise in entropy will take place as the Universe tends to a thermodynamic equilibrium. This could seemingly be true as no polymer polymerizes endlessly to attain infinite molecular weights.

If our planet Earth is mimicking the laws ruling our Universe, then every polymerization process should go to end whether controlled by genetics or other factors, such as lack of proper conditions. However, stopping a polymer from further polymerization does not mean the death of the polymerized object as it should first pass through an aging or deterioration process. If so, are the observed black holes a manifestation of an aging process of our Universe? Well, not necessarily since substantial data indicate that the Universe is still expanding. Thus, perhaps the rarely formed black holes are an indication of the death of some tiny spots of the Universe as part of the growing process. Living organisms on Earth experience cell death and regeneration during growing processes but once aging takes place, cell death becomes recurring until total deterioration.

Whether the Universe is still polymerizing to finalizing its shape or started an aging process, what role does humanity play in all of this? In other words, are humans appeared by fortunate accident as a result of suitable conditions or they are created for a specific goal by some sort of extraterritorial or ''extra-universal'' civilization? Well, the answer may depend on whether humans are significant and play a crucial role in the function of this vast Universe. Humanity occupies a very tiny place in the Universe, equivalent to some dirt deposited on the surface of one single electron in a tree. If we travel at the speed of light (186 000 miles per second), it will take at least 100 000 years to only cross our Milky Way galaxy. And as mentioned previously, the Milky Way is just one galaxy among at least billions of other galaxies. Thus, we occupy a negligible place in Universe less than a grain of sand lost in the whole Sahara desert. If our observable Universe formed 13.8 billion years ago, by shrinking down the span of time to 24h, with the Big Bang event happen at 12 am, the first Homo sapiens appeared just a few minutes before the following 12 am.

Numerous studies in the framework of abiogenesis propose that chemicals encountered in non-life have the opportunity to naturally combine and transform in a way to produce life. This conversion from non-life to life may not be noticeable to the naked eye as it takes extremely long periods of time while complexity progressively rises[33]. As a result, numerous researchers make use of several disciplines, such as chemistry and paleontology, in an effort to extrapolate and identify the very early Earth's conditions and pre-life chemical processes that took place billions of years ago to generate life. The biochemistry of life may have commenced 13.8 billion years ago, just after the Big Bang during the inflationary epoch or time when the Universe was yet 10 to 17 million years old. Some argued that life begins with particular chemicals involving carbon and water building four key families of chemicals, namely lipids (fatty cell walls), carbohydrates (sugars, cellulose), amino acids (protein metabolism), and nucleic acids (self-replicating DNA and RNA). Relevant interpretations of the origins and interactions between these chemicals contributed to the success of several abiogenesis theories. The first idea proposed that life on the planet Earth may originate from RNA-based life. However, other lifeforms existing prior to RNA must be considered.

According to the Miller-Urey experiment[34], amino acids as forming blocks of proteins present in living organisms would have the potential to be produced by combining various inorganic compounds. However, this experiment should satisfy the proper conditions present in early Earth, which will enable replication and polymerization to develop higher complex systems. This might be sparked by external factors like lightning and irradiation offered by early Earth's conditions. In the Metabolism-first hypothesis, chemical systems of early Earth may have been catalyzed in a way to create precursor molecules. In other words, certain catalytic processes brought to life the primary molecules able to self-replicate and polymerize into complex arrangements, also known as "spontaneous generation" in evolution. Our solar system and the vast interstellar space provided the material that may have contributed to provoking the development of life on Earth.

Another study known as Panspermia Hypothesis offers an alternate interpretation of the existence of life on Earth[35,36]. It proposes that microscopic life is imported from elsewhere in the Universe by space dust, meteoroids, asteroids, and small solar system bodies. This implies that life may be dispersed all around the Universe and occurred outside Earth and not on Earth itself.

So far, most studies concluded that Earth is the only planet accommodating life. However, future advanced technologies may prove otherwise. Fossil evidence further confirmed the abiogenesis claims, where over 99% or almost five billion living species ever resided Earth's surface have now become extinct. The oldest discovered uncontested proof of life dates back to at least 3.5 billion years ago, just after solidification of geological crusts. In May 2017, evidence of early life dating back to 3.48 billion-year-old was revealed in Pilbara Craton, Western Australia. Other mineral deposits were also identified in geysers and hot springs, supporting the existence of much earlier life forms on Earth. Microfossils located in hydrothermal vent precipitates dating back to around 3.77

to 4.28 billion years old were excavated in Quebec, Canada[33]. These were acknowledged as the oldest record of life on Earth, proposing "an almost instantaneous emergence of life" after formation of oceans 4.4 billions of years ago.

If the Universe comes from one single point in time, then life should adhere to the same principle. Although the living organisms crawling our planet Earth look distinct from one another in many aspects varying from body parts to skeletons and exoskeletons aspects, they all carry the same origin. Biologists do believe that those attributes were once the same and later evolved to higher forms and different species. During this process known as "evolution"[37], the attributes improved to increase the survival chance. Consequently, some species have gone extinct during the evolution process and others made it to the top of the food chain. In other words, everything evolved over time from the common ancestral gene pool to higher forms. This common ancestor appeared more than 3.5 billion years ago, and the genetic code held by the ecosystem is a solid argument of the universal descendent of all bacteria, archaea, and eukaryotes.

So, how did this idea of evolving from one life form to another all started? Well, it did not appear in the era of the Big Bang theory but first introduced by Jean-Baptiste de Lamarck in 1809. More than 50 years later, the concept gained greater strength when Charles Darwin and Alfred Russel Wallace proposed that nature´s driving force to evolution is natural selection[37]. In other words, some species prosper and some perish depending on processes subject to selective breeding. If so, is there a controlling hand within nature to favor which of them will survive or die? Not necessarily but nature itself can dictate the species able to survive by adapting and evolving to higher forms. Modern studies did found that genetic drifts do occur during evolution, where some genes present in one category of species vanish if they cannot survive or breed, further strengthening the evolution theory.

If basic chemical elements exposed to early Earth's specific conditions could create life and evolve to diverse other lifeforms then why related processes could not regulate our Universe at first. After all, the Universe formed much earlier than life on Earth. So, what if life on Earth is just leftovers of collective memory processes governing our Universe? For sure comprehending such things from our infinitesimal place in the Universe will be inconceivable. Some dirt deposited on one single electron in a tree will merely see its neighboring electrons, nucleus, and other nearby atoms or molecules, similar to planets, stars and galaxies perceived by bare eyes or telescopes. However, it will be impossible for it to envisage that the tree is a fully living intelligent organism, grown virtually from nothing (the seed) to form something at the end the process, and then deteriorate through aging to provide opportunities to other organisms to develop. Each of these electrons, atoms and molecules are in perpetual motion, have energies, emit or absorb irradiation, just like planets, stars and galaxies. So what if our Universe is an organism with a distinct shape and maybe intelligence just like a tree, an animal, a human body or something else! Note that the term intelligence is employed to express the ability of any living organism to solve problems. Humans are not the sole species capable of solving problems but animals and plants also manage in their own ways in order to survive.

If our Universe is so vast, and we are so insignificant, does this mean that we are unimportant in the full picture of this vast Universe? Well, because we are infinitesimally small, this gives us the sense of insignificance. However, the significance may also be in function of value. If planet Earth is equivalent to an electron, then as I said we are just some dirt deposited on its surface, which may or may not affect the properties of the electron. Certainly, electrons play significant roles in electricity, magnetism, thermal conductivity and chemistry but our technologies are so far inadequate to differentiate between electrons and visualize the surface of each of them. It is possible that modification of the electron surface will affect its properties but this still has to be confirmed. On the other hand, changing the surface of one electron in a tree will presumably not impact the whole tree, unless there is a recurring pattern, meaning that a significant number of electrons are altered. This implies that other intelligent lifeforms like us are spread throughout the Universe.

Whether or not we contribute to the operation of our Universe, should our cosmic insignificant lead us to despair and unvalue? Not at all since concepts of value, whether associated with happiness and/or material means, lay on our planet Earth. In our life spam, whether or not these concepts are objectively valuable, it only matters to each of us and do not rest on having the command over our grand Universe. Our responsibility is to treat one another, as well as our surrounding, with respect and consideration.

References

(1)   Zeilik, M.; Gregory, S. A.; Introductory Astronomy and Astrophysics (4th ed.)., Saunders College Publishing, 1998.

(2)   Woolfson, M. The origin and evolution of the solar system, Astronomy and Geophysics. 41(1), 2000, page 12.

(3)   "Star system" in Modern Dictionary of Astronomy and Space Technology. A.S. Bhatia, ed. New Delhi: Deep and Deep Publications, 2005.

(4)   Croswell, K., Planet Quest: The Epic Discovery of Alien Solar Systems. New York: The Free Press, page 43, 1997.

(5)   Birth of the Universe, University of Oregon, 2016 (discusses "Planck time" and "Planck era" at the very beginning of the Universe).

(6)   Evans, D.S., Stars of Higher Multiplicity, Quarterly Journal of the Royal Astronomical Society, 9, 1968, pages 388–400.

(7)   Singh, S., Big Bang: The Origin of the Universe, Harper Perennial, 2005, page 560.

(8)   Big bang theory is introduced, A Science Odyssey, WGBH, 1927.

(9)   Smoot Group, The Cosmic Microwave Background Radiation, Lawrence Berkeley Lab, 28 March 1996.

(10) Loeb, A., The Habitable Epoch of the Early Universe, International Journal of Astrobiology, 13 (04), 2014, pages 337–339.

(11) Guth, Phase transitions in the very early Universe, in: Hawking, Gibbon, Siklos (eds.), The Very Early Universe, 1985.

(12) Ryden, B., Introduction to Cosmology, Addison-Wesley, 2003, page 196.

(13) Overbye, D., Cosmos Controversy: The Universe Is Expanding, but How Fast?, New York Times, 2017.

(14) Pullman, B., The Atom in the History of Human Thought, Oxford, England: Oxford University Press, 1998, pages 31–33.

(15) Stern, D. P., The Atomic Nucleus and Bohr's Early Model of the Atom, NASA/Goddard Space Flight Center, May 2005.

(16) Fackler, O.; Tran, J. T. V., 5th Force Neutrino Physics, Atlantica Séguier Frontières, 1988.

(17) Buchwald, J. Z.; Warwick, A., Histories of the Electron. MIT Press, 2001.

(18) Lewis, G. N. The Atom and the Molecule, Journal of the American Chemical Society, 38 (4), 1916, page 772.

(19) Atkins, P.; Loretta, J., Chemistry: Molecules, Matter, and Change. New York: W. H. Freeman & Co. 1997, pages 294–295.

(20) Young, R. J. Introduction to Polymers, Chapman & Hall, 1987.

(21) Mohanty, A. K.; et al., Natural Fibers, Biopolymers, and Biocomposites, CRC Press, 2005.

(22) Kumar, A.; et al., Smart Polymers: Physical Forms & Bioengineering Applications, Progress in Polymer Science, 32, 2007, page 1205.

(23) Fletcher, J., On the functions of organized beings, and their arrangement, In: Rudiments of physiology. Part 2, On life, as manifested in irritation, Edinburgh: John Carfrae & Son., 1837, pages 1-15.

(24) Rubinstein, M., Colby, R. H., Polymer physics. Oxford, New York: Oxford University Press, 2003, page 6.

(25) Yu, E. H.; Snadmacher, K.; Enzyme Electrodes for Glucose Oxidation by Electropolymerization of Pyrrole, Trans IChemE, Part B, Process Safety and Environmental Protection, 85(B5), 2007, pages 489–493.

(26) Ammam, M.; Fransaer, J.; Microbiofuel cell powered by glucose/O_2 based on electrodeposition of enzyme, conducting polymer and redox mediators. Part II:

Influence of the electropolymerized monomer on the output power density and stability, Electrochimica Acta. 121, 2014, pages 83-92.

(27) Trimble, V. Existence and nature of dark matter in the Universe, Annual Review of Astronomy and Astrophysics, 25, 1987, pages 425–472.

(28) Peebles, P. J. E.; Ratra, B., The cosmological constant and dark energy, Reviews of Modern Physics, 75 (2), 2003, pages 559–606.

(29) Fred C. A.; Gregory, L. A dying Universe: the long-term fate and evolution of astrophysical objects, Reviews of Modern Physics, 69 (2), 1997, pages 337–372.

(30) Penrose, R., The Road to Reality: A Complete Guide to the Laws of the Universe, Vintage Books, 2005.

(31) Hawking, S., A Brief History of Time, Bantam Books, 1988.

(32) Brush, S. G. A History of Modern Planetary Physics: Nebulous Earth, Cambridge University Press, 1996, page 77.

(33) Dodd, M.. et al., Evidence for early life in Earth's oldest hydrothermal vent precipitates, Nature, 543 (7643), 2017, pages 60–64.

(34) Hill, H. G.; Nuth, J. A. The catalytic potential of cosmic dust: implications for prebiotic chemistry in the

solar nebula and other protoplanetary systems, Astrobiology, 3 (2), 2003, pages 291–304.

(35) Berera, A. Space dust collisions as a planetary escape mechanism, 2017.

(36) Chan, Q. H. S., et al. Organic matter in extraterrestrial water-bearing salt crystals, Science Advances. 4 (1), eaao3521, 2018.

(37) Coyne, J. A., Why Evolution is True. Oxford: Oxford University Press, 2009, page 17.

www.ingramcontent.com/pod-product-compliance
Lightning Source LLC
Chambersburg PA
CBHW030040230526
45472CB00002B/596